Qu

Physics For

Beginners:

Quantum Mechanics and Quantum Theory Explained

By

Jason Stephenson

Table of Contents

Quantum Physics For Beginners: Quantum Mechanics and Quantum Theory Explained

By Jason Stephenson

First Published, 2015

Printed in the United States of America

Introduction

We may have lived knowing that the world around us operates in a way as if we observe them to be. This knowledge of how the universe operates, based primarily of our observations, has enabled us to predict actions and motions and allowed us to build machines and equipments that have made our lives easier and more enjoyable.

The field that allowed us to do that is classical physics.

The world, however, is advancing and our knowledge of how things are expands over time. We have discovered in the last few decades that these sets of rules that we have devised can perfectly describe the large-scale world but cannot accurately define the behaviors of particles in the microscopic world. This necessitated another field to explain the different behavior in the microscopic world: quantum physics.

Knowledge of the rules that govern the operation in the large-scale is important, but it turns out that what happens in the microscopic world greatly defines the outcomes in the macroscopic world.

This book simplifies Quantum Physics for beginners and for those who would like to explore the subject more. In here are basic concepts and application in the field for you to be more acquainted with the subject.

Chapter 1. Quantum Physics Defined

Matter and the order of things we deal with and observe every day follows a set of rules and patterns that we can understand: objects fall upon the earth due to gravity, speeding objects move at a straight path at a constant speed unless acted upon by a force, etc., Those rules govern not just the things around us but the entirety of this realm we call the universe. However, those rules affect only the large things in this world (macroscopic world). If we observe how little things on the microscopic world behave, we will find sets of rules different to those that govern the visible world. Atoms and microscopic particle, for example, are not affected by gravity.

For the tiny things that constitute matter and the everyday object we encounter, we need a different discipline that will focus on the rules that govern them.

Quantum physics (also known as quantum mechanics and quantum theory) is the study of the physical phenomena occurring at a microscopic scale. On this scale, the classical laws of physics cannot define the behaviors of particles. Quantum Physics Defined

Matter and the order of things we deal with and observe every day follows a set of rules and patterns that we can understand: objects fall upon the earth due to gravity, speeding objects move at a straight path at a constant speed unless acted upon by a force, etc., Those rules govern not just the things around us but the entirety of this realm we call the universe. However, those rules affect only the large things in this world (macroscopic world). If we observe how little things on the microscopic world behave, we will find sets of rules different to those that govern the visible world. Atoms and microscopic particle, for example, are not affected by gravity.

For the tiny things that constitute matter and the everyday object we encounter, we need a different discipline that will focus on the rules that govern them.

Quantum physics (also known as quantum mechanics and quantum theory) is the study of the physical phenomena occurring at a microscopic scale. On this scale, the classical laws of physics cannot define the behaviors of particles.

The Mystery of the Black Body

Before the birth of quantum physics, classical physics was bothered by the problem in the radiation of blackbodies.

Objects emanate light when subjected to high temperatures. This is because heat adds up to the kinetic energy of electrons on the material. Electrons tend to give off light whenever their speed is increased or decreased.

Despite many who attempted to solve the black body problem, nobody succeeded. It turned out that the problem is due to energies being quantized. Max Planck, one of the pioneers of the field, theorized that a black body does not emit a continuous stream of light when heated but emits light in discrete quantities called quantum (or quanta). After Max Planck converted the formula created by Raleigh-Jeans to describe the behavior of light into a form where light is a sum of an infinite number of discrete quantities. Only after then that, the formula could describe the behavior of light correctly.

Photoelectric effect

This is another physical phenomenon, which is a result of light being quantized called the photoelectric effect. When a piece of metal is exposed to light, electrons are released. Light is interpreted in classical physics as a wave of electromagnetic radiation. This wave has amplitude and frequency, similar to other waves, and that its amplitude correlates to its frequency. Classical physics explant the release of electron as the effect of the wave. The higher the amplitude (intensity) of the wave, the more it is capable of dislodging an electron. However, experiments failed to prove that. In fact, there are instances where no electron is released, however, intense the light is. Furthermore, the kinetic energy of the electron released was found to be independent of how intense the radiation used was. It is only when Einstein proposed that light travels in discrete quantities that photoelectric effect was finally correctly explained.

Positrons

This positively charged version of electron was first postulated and proposed by Paul Dirac. Before quantum physics, the idea of particles of opposite charge (like a positively charged electron) is outlandish. As proof to the particle-wave duality of light as proposed by quantum physics, scientists have discovered that a photon of light can be split into two oppositely charged particle. Later, they discovered that two anti-particles annihilate each other producing a photon of light. Later, this was demonstrated by Einstein's theory of Relativity in his famous $E=mc^2$ equation.

The Dual Nature of Particles and Waves

While scientists were gradually being convinced that waves exhibit particle-like properties, Louis de Broglie suggested in 1923 that particles also exhibit wave-like properties.

In a PhD thesis published by Louis de Broglie long before Quantum Physics was developed, he established an equation describing the wave-particle properties of matter, that is $\lambda=h/p$ and $k=p/h$.

Although an a particle, say an electron, displays wave-like and particle-like properties, it does not mean that they appear physical both a wave and particle as well.

The duality of their nature is only exhibited on the *measurement* aspect of observing them.

Playing Dice with Gods

Many scientists, including Albert Einstein is not comfortable with Quantum Mechanics simply because the field is filled with uncertain numbers.

In classical physics, objects can be described using other values as momentum and position. Both values can be measured exactly. This is a proof of classical physics being extremely deterministic in nature.

On the microscopic world, however, tiny particles behave differently. You cannot pinpoint their exact location because the values that could describe their motion are not directly measurable. For this, mathematics and science have produced a different set of equation. One of the principles that emerged that could describe this is the Heisenberg Uncertainty Principle. It states that the momentum and position of elementary particles cannot be precisely measured and, hence, exhibit an inherent degree of uncertainty.

The principle is described as

$\Delta x \Delta p_x \geq \hbar/2$

Where Δx is the uncertainty in the position of the particle, Δp_x is the uncertain in the momentum (at x) and $\hbar = h/2\pi$. This equation states that we cannot determine, exactly, both the position of the particle and its momentum. We can only determine the other accurately while leaving the other more inaccurate. In order to describe both, once must give probabilities of the values for momentum and position. This uncertainty principle is a problem inherent in the duality of the nature of matter since the location of waves cannot be determined exactly in space.

A wave function, $\psi(r, t)$, describes the state of particles. Squaring the wave function yields the intensity of the wave function (i.e. $|\psi(r, t)|$). In here, the amplitude of the wave serves as the function of the position and time for particles. Quantum physics tell the position of a particle through its *probability density* computed by converting the position and momentum into probabilities in wave-function. The momentum-space version (contrary to the above position-space version) of the wave function is $\varphi(p, t)$.

Chapter 2. Applications of Quantum Physics

Quantum physics is mysterious and weird. Though it accurately describes the mechanisms of microscopic bodies, still, it defies human logic pertaining to realism and causality.

Among weirdest of the laws of quantum physics is entanglement. The rule says that electrons on the opposite side of the Universe (the farthest distance possible between two electrons) can be linked and share information instantly. However, in the macroscopic scale where classical law of Physics applies, this cannot be possible. In fact, all things are bound by the speed limit of the Universe, which is the speed of light.

Quantum physics is riddled by many rules that are not similar to the rules of the classical physics. The lists of application for those are limitless. Among the applications for those are:

Ultra-precise Clocks. Clocks are important part of everyday life. Most clocks, however, are not precise, as they tend to lose seconds over time making them inaccurate keepers of time over a number of years. The first clock ever built, the quantum-logic clock, uses the principles of quantum mechanics—it measures the radiation frequency required to make electrons move from one energy level to another. The said clock is housed in the U.S. National Institute of Standards and Technology in Colorado and is known to have a degree of error of plus or minus 1 second over a period of 3.7 billion years. The most precise one, the NIST strontium clock, is said never to gain or lose a second for up to 5 billion years.

Uncrackable Encryption

In conventional encryption, codes can be intercepted during its journey from the sender to the recipient. This encryption is subject to be being intercepted and decoded. Using quantum physics, though, a new type of encryption known as quantum key distribution is developed. In this kind of encryption, the information about the key is sent through a randomly polarized photon (a photon is a quanta of light) to restrict its vibration to only one plane (i.e. horizontal or vertical). The recipient then use filters and algorithm is used to decrypt the key. This technology is now being used by various companies such as ID Quantique, Toshiba and BBN Technologies. ID Quantique developed a system of voting that is tamper proof for an election. This technology is also now being used to secure the bank transfers from as early as 2004.

Biological Compasses

Although this technology is yet to be developed and discovered, scientists have observed this incidence on birds and believe that the ability of the birds to migrate from a place to another without any aide is due to the principle of quantum entanglement. Scientists believe that a certain form of molecule known as cryptochrome are found in the eyes of the birds which allow them to *see* the earth's magnetic field via entangled pair of electrons.

Powerful Computing Machines

Conventional computers store and process data using a pair of binary digits (0, 1). Quantum computers, however, take advantage of superposition and stores data in the form of quantum bits or qubits. These data are either 1s or 0s until the data is interpreted or evaluated. This computer is said to have many times the power of conventional computers. The first company to have built this is D-wave with its first 128-bit and 512-bit processors used and tested today by NASA and Google Laboratories namely D-Wave one and D-Wave-two.

Improved microscopy

Using entangled pairs of photons, scientists from the Hokkaido University in Japan have been able to improve the resolution of microscopic images and call the newly formed device entanglement-enhanced microscope.

Chapter 3. Dealing With Probabilities

In classical physics, your goal is to arrive at precise and exact measurements—you use exact numbers to describe your phenomenon and use them to arrive at answers. In quantum physics, however, the numbers you are mostly likely to deal with are probabilities—range of probably numbers that *could* define your phenomenon with the highest degree of accuracy.

Vectors in Hilbert Space

Named after David Hilbert, the Hilbert space is a hypothetical space, which extends the two-dimensional methods of calculus and vector algebra.

Say for example you have the following values:

2	1
3	2
4	3
5	4
6	5
7	6
8	5
9	4
10	3
11	2
12	1

The first column is the sum of the rolled dice with the second column is the relative probability of rolling them. Thus, you can twice as likely to roll a seven than a four, for example.

In order to do this, you have to arrange them in vectors:

[1]
[2]
[3]
[4]
[5]
[6]
[5]
[4]
[3]
[2]
[1]

Now, the vectors in the column above state the relative probability of rolling dice. In quantum physics, however, probabilities are not directly used. Instead, the square roots of them called the *probability amplitudes*. The actual probability, then, is determined by using the wave functions and then squaring them afterwards. Thus, the first step is to get the square root of the function is:

$$\begin{bmatrix}\sqrt{1}\\ \sqrt{2}\\ \sqrt{3}\\ \sqrt{4}\\ \sqrt{5}\\ \sqrt{6}\\ \sqrt{5}\\ \sqrt{4}\\ \sqrt{3}\\ \sqrt{2}\\ \sqrt{1}\end{bmatrix}$$

The next step is to divide the square roots by the root total of the original numbers (i.e. 36). Thus,

[√1/6]
[√2/6]
[√3/6]
[√4/6]
[√5/6]
[√6/6]
[√5/6]
[√4/6]
[√3/6]
[√2/6]
[√1/6]

Thus, based on our results, the probability of rolling a 2 is √1/6, of 3 is √2/6, and so on...

Dirac Notation

The technique outlined above is only good on simple possible states a particle can assume as with the dice space. Most of the problems dealt with by quantum physics in the real world, however, has vectors that are extremely (often, infinitely) gigantic. Using vector notations to handle this is not easy, thus another notation, created by and named after Paul Dirac, was created—the *bra-ket notation or the Dirac notation.*

In diracts notation, vector notations can be abbreativated in kets. The above notation can be written as this:

$$
\begin{array}{ll}
 & [\sqrt{1/6}] \\
 & [\sqrt{2/6}] \\
 & [\sqrt{3/6}] \\
 & [\sqrt{4/6}] \\
 & [\sqrt{5/6}] \\
|\psi > & [\sqrt{6/6}] \\
 & [\sqrt{5/6}] \\
 & [\sqrt{4/6}] \\
 & [\sqrt{3/6}] \\
 & [\sqrt{2/6}] \\
 & [\sqrt{1/6}]
\end{array}
$$

Operators Used in Quantum Physics

Hamiltonian (H) – This operator corresponds to the total energy (E) in a given system. It is represented by $H|\psi> = E|\psi>$.

Gradient (∇) – generalizes the mathematical concept of the derivative:

$\nabla|\psi> = \partial/\partial x\ |\psi> \boldsymbol{I} + \partial/\partial y\ |\psi> \boldsymbol{j} + \partial/\partial z\ |\psi> \boldsymbol{k}$

Unity or identity (I): This operator does not change kets:

$I\ |\psi>= |\psi>$

Linear Momentum (P): $P\ |\psi>= i\hbar\nabla\ |\psi>$

Laplacian (∇^2) – used for the creation of Hamiltonian operator that is an energy-finding, Laplacian is a second-order gradient:

$\nabla^2|\psi> \nabla.\nabla\ |\psi> = \partial^2/\partial x^2\ |\psi> + \partial^2/\partial y^2\ |\psi> + \partial^2/\partial z^2\ |\psi>$

Chapter 4. Particles in Bound States

A particle in a bound state is that which cannot move and tends to remain in one or more regions of space. This may be due to an external factor or the presence of another particle which binds the said particle. A particle, for example, that is trapped in a square well may be likened to a pea inside a box which cannot move elsewhere but only on the space bound by the box. While classical physics may find it difficult to deal with those, quantum physics excell at those.

The equation used to describe the moment is the Schrödinger equation:

$$-h^2/2m\, \nabla^2 \psi\,(r) + V(r)\,\psi(r) = E\,\psi(r)$$

The square well

The square well, also known as the particle in a box model, is a hypothetical box used to demonstrate how classical physics is different from quantum physics. In a classical physics setup, for example, a ball may represent the particle and the box may represent the bounding energy. The ball inside the box could be at any place, could assume a specific amount of energy or speed and can have a state of zero energy (or a state where a ball stand still). Reducing the size of the particle, however, encourages a different scenario. That is, particles tend to follow the quantum law more instead of the physical law. The said particle, for example, could only occupy certain positive state and that particles can never have zero energy. Quantum mechanics also provide that there are areas in space that a particle could have more probability of residing than other areas.

Without sufficient energy to break free, there is a tendency for a particle to be bound in a potential well.

The total energy of a particle is the sum of its potential and kinetic energy. This sum is always constant:

$$p^2/2m + V = E$$

If the total energy of the system is less than its initial V, then a particle could be trapped in the well. Analyzing the equation, we could arrive at a solution that in order for the particle to escape the bounding potential box, it needs to assume a negative kinetic energy, which, as what we have learned so far and as what human capability permits, is highly impossible.

Bound state

A particle is said to be in a bound state when it is restricted to travel freely into the infinite space—simply put, when it is stuck in a potential well. The bound states are found to form a discrete spectrum of energy levels in which the particle is only allowed to reside.

As what we have learned from the formula stated earlier, a particle is bound in a potential well if its energy *(E)* is less than the sum of both the initial and final V (i.e. V_1 and V_2). The behavior of such particle can be described by the Schrödinger equation for which the energy states and wave function that a particle can assume is determined. In order to solve the problem completely, two boundary conditions must be solved. These two boundary conditions are:

Initial V is less than the total energy, which is less than the final V ($V_1 < E < V_2$). The particles in this condition have sufficient energy to overcome the bounding energy in the left while not having enough to overcome the bounding potential energy in the right. Thus, the particle is allowed to move on the left side, which is the negative region of the x-axis (i.e. region between $-\infty$ and x1).

Because the region to the left of the bounding potential box is open, the allowed energy values are not discrete but are rather continuous.

Energy (E) is greater than final V

In this state, the particle is not bound and is free to travel to both positive and negative infinities.

Chapter 5. Harmonic Oscillators

In physics, these oscillators are setups with periodic motion such as the pendulums and bouncing springs. There are already provisions and rules governing the motion of oscillating bodies in the macroscopic world however, the conditions when it comes to microscopic scales become different.

Harmonic oscillation could usually occur in the bonds of the atoms in a molecule. These bonds are not fixed and are therefore subject to oscillation as if they are springs able to move back and forth.

In classical physics, harmonic oscillation is described by the Hooke's law:

$F = -kx$ *(1)* where k refers to the spring constant and x is its displacement. In classical physics, the force needed to restore an oscillating body to original state is directly proportional to the length of oscillation—i.e. the farther the spring is pulled, the greater force is required to pull it back.

Recall that in classical physics, the force of a body is the product of its mass and its acceleration (*F=ma)*, thus, rewriting the above-mentioned formula gives us:

$ma=-kx$ or $ma+kx=0$ (2)

The equation for instantaneous acceleration:

$a = d^2x/dt^2$

Rewriting equation (2) gives

$m(d^2x/dt^2) + kx = 0$

Classical physics provide that the quotient of the constant and the mass equate to the square of the angular frequency, thus

$d^2x/dt^2 + \omega^2 x = 0$

For values of x at a constants A and B values:

$x=A\sin \omega t + B\cos \omega t$

The sin and cos represent that the waveforms are of oscillation body since it is periodic.

In classical physics, the total energy of a system is the sum of its potential and kinetic energies

H = KE + PE

For a particle, its kinetic energy is defined by:

KE=$p^2/2m$ where p is the momentum of a particle and m is its mass.

The potential energy of the particle is:

$$PE = ½\, kx^2 = ½\, m\, \omega^2 x^2$$

Thus, the Hamiltonian equation mentioned above can be rewritten as follows:

$$H = p^2/2m + ½\, m\, \omega^2 x^2$$

Angular momentum

Angular momentum in the microscopic world occurs more than the usual golf-ball-attached-to-a-string setup in the macroscopic world whirling above your head. Imagine two atoms held together by a bond and spinning in space.

Aside from having both potential and kinetic energy, a particle exhibits another form of energy: *rotational energy.*

In $H=L^2/2I$ where L is the angular momentum, I is the inertial moment of rotation.

Spin

Experts agree that the angular momentum exhibited by atoms in a molecule is not the only momentum it contains. Another form of momentum called *spin* is exhibited by electrons. This spin is a mathematical description. Whether electrons actually spin (as if in an imaginary axis) or not can never be determined. What is certain is that *electrons have intrinsic angular momentum.*

Chapter 6. The Future of Quantum Physics

Harnessing the power of the law of quantum dynamics has always given birth to technologies that greatly surpass our present technology. The following technology will likely be developed in and will benefit the future:

Quantum Dots

Imagine a solar panel so small it could even fit into your purse yet so powerful it could be used to power all your devices at once. This may sound implausible in the present but the future holds promise of the next generation of solar panels that are able to take benefit of quantum dots—tiny bits of semiconductor .

The size of these semiconductors is designed in such a way that electrons only reside at specific areas for maximum absorbance of light and of energy as well.

Quantum Wires

The problem with the usual wires of today is that they are wasteful when it comes to energy. They have resistance, which converts some of the electrical energy to thermal energy, which is expelled in the environment. Quantum wires, however, are wires made so small that the electrons are made to move in a specific way thus making it a very good conductor of electricity.

Superconductors

In addition to superconducting wires, one of the possible feat of quantum mechanics in the future is superconductors. In this material, atoms are aligned in such a way that the electron flow is maximized thereby minimizing resistance.

Light Emitting Diodes

The present technology which benefits quantum mechanics and could improve in the future is LED's. This revolutionary incandescent light-producing contrivance is efficient and is replacing fluorescent lamp as we speak.

The future of quantum mechanics holds much promise of more technologies with far-reaching effects. Quantum mechanics is at its infancy now yet our present technologies are greatly influenced by it.

Conclusion

What you just read now comprise your basic knowledge of the complex world of quantum physics. The concepts outlined herein should have given you a means to explore the world of quantum mechanics further. Thus, your journey does not end here. To be able to know more and to have a deeper grasp of the field, you need to continue reading.

What you have learned here somehow represents that entirety of what quantum mechanics is concerned about. To be able to advance into it more, you have to continue with the next step: taking a class on the subject.

Thank You Page

I want to personally thank you for reading my book. I hope you found information in this book useful and I would be very grateful if you could leave your honest review about this book. I certainly want to thank you in advance for doing this.

If you have the time, you can check my other books too.

Lightning Source UK Ltd.
Milton Keynes UK
UKHW011120301121
394802UK00008B/1600